ON THE PRECIPICE
with
Gravity, Levity and Vertigo

by
Norman E. Dunkle

ISBN 0-936369-79-5
Cover design by Country Lane Graphics, New Castle, PA
(with detail from an original wood carving by the author)

SON-RISE

143 Greenfield Road
New Wilmington, PA 16142
1-800-358-0777

Printed in the USA

TO LUCILLE,
of course

IN GRATITUDE

I am indebted to many people for their encouragement to prepare this collection for publication. Above all, I'm grateful to my wife, Lucille - my best friend and partner for nearly forty-eight years.

I also want to thank the members of the Writers' Roundtable of Mercer, Pennsylvania - especially their leader, Evelyn Minshull, and my publisher, Florence Biros. That group very literally encourages (imparted courage) at times when I was hesitant to see this material through to publication.

I'm especially grateful to Andrew Stallsmith, a colleague in ministry and poetry, who introduced me to the Writer's Roundtable. Andy not only encouraged, but led by example, shattering all my negative stereotypes of preacher-poets.

Norm Dunkle

CONTENTS

PROLOGUE

THE PRECIPICE

I clutch and cleave to this precipitous face
of logic, solid reason -
Suspended here above wild vacant space
and free-fall madness, treason
Against the sovereignty of mind and grace.

Edging past alarm, I hang inured,
Constrained, impeded, but likewise secured
by ponderous gravity
and drawn by heady hope
that gravity is Truth,
and levity is Wisdom.

* * *

I'll rest here then,
on this slim rim of thin conjecture,
bare edge of solid inference,
A slender ledge of logic to
secure my dubious balance -
capricious equipoise -
above the hungry chasm of
brute chaos gaping there below.

Fatigued, I shift and slip
then grasp a brief reprieve.
But still remote from Wisdom,
who beckons at the top
of this sheer precipice,
I Fumble blindly overhead.
I stretch and grope to touch some hope,

some hold on speculation,
Then peer above and strain
to penetrate the mist.
But throat constricts.
Neck aches and cramps.
My head drops down
To face . . . the dread below
And ponder my dilemma:

I can't stay here,
constrained in fear,
But neither can I climb
until I loose my grip on these,
redundant old securities
and reach for higher mysteries.

The sheer hard facts remind me,
it's only past behind me.
The quick way back is death.
So risk! I'll risk
The storms of honest doubt,
the gale of Wisdom's breath,
And Truth's portending wrench,
to hard reality.

I grope and scratch to left and right.
Ahh, here's a narrow paradox,
an aperture between two rocks
of contradiction,
Where I can wedge my wavering mind,
jammed tight, constrained,
secured - time gained.
But yet I dare not . . .
Dare not rest my weight here long,
where Truth complains and cries

for evidence.
I must induce, deduce
some sense for me,
some sensory
support on which to think
toward that distant brink.

And there's a clear though slender ledge,
a syllogisms' cutting edge.
I saw it briefly; now it's slipped
away, just out of sight,
an inch or two beyond my grasp.
I stretch and claw then lunge and gasp
for fear - and hope.

Elusive hope betrays and slips
away from under finger tips,
evades, defies my zealous quest,
and laughs when I exert my best.

* * *

But still it's clear, I can't stay here:
I'll clamber on and trust
(But trust. . . in what?)
While winking Wisdom waits,
remotely levitates
in lofty contemplation,
alluring, but elusive;
And ominous Truth is here
and everywhere -
salvation if I trust,
my death if I defy,
and torment
if I linger undecided.

* * *

I'll trust the truth of gravity
and gravity of truth.
Sheer mass itself,
its massive hold,
And weighty universal force,
insists, resists,
Secures me to this ledge,
Yet threatens to annihilate.

By weight, the heft, the heaviness,
import, support of mass,
It's possible to climb or fall,
and I must choose my way.

But, God, please let me see and help me heed
Transcendent Verity and hear him say,
"I am the Truth,
I am the Life,
the Way."
Such Truth is no mere tribute earned,
no pinnacle I strive to gain
to plant a flag that may remain
while I descend.
But here is Truth personified,
Omnipresent, unified.
Nor can I grasp such Truth, but he
is present stark reality,
and in my peril Truth grasps me.

Good Truth, please hold me 'til I find
Some higher place to rest my mind,
Then grant me Wisdom's levity.

I.
THE SOUND OF CLASHING CULTURES

"Weakness is being lied into something meritorious, no doubt of it. . . ."*

Friedrich Nietzsche,
On The Genealogy of Morals

"The cheap promise of radical newness is the most boring and repetitious of all modern ideas."

Thomas Oden,
Agenda for Theology

* The object of Nietzsche's scorn was, of course, traditional morality; but his insight is always more incisive when we let it cut both ways like the penetration of a double edged sword - especially now when every form of immorality seems resolved to redefine its weakness as a virtue.

TO POETRY

Poetry! Old friend!
Musing confidant of childhood.
Vowel taster, word waster,
Connoisseur of verbal melodies
And cadenced heartbeat vortices.

Forgive me if I seem suspicious, though.
We've hardly spoken now for forty years -
A joke at most, a wink, a meaningless flirtation,
Children's song or limerick,
Terse verse to tease or please
A grandchild or a friend.

And how you've changed!
I've heard the news -
How fools abuse
You, cherished Muse.
Hostile minds who never knew you
Flush their foul infection through you.
A fevered culture's refuse overruns
with ignorance
and arrogance,
cynics' discontent,
vulgar sophistry,
raving madness,
frenzy's pitch and tempo.
For decades now we knew that you'd
Been misperceived and misconstrued
By those who hate and violate
Your sister Muses also.

Of course, you always had a bawdy side -
Hackneyed verse, for fun or ridicule.

But now they've covered you in cheap contempt,
And cheaper noise -
Bedlam's clamor, driven rhythm,
Shrieking madhouse vocalize,
Infernal loathing, wild malevolence.
And you, poor Muse, you play the fool,
Loquacious imbecile,
Eager entertainer to
That fevered brood of insolence,
Wild fury's mad debauchery,
And vulgar, raving violence.

* * * *

Oh! You protest; it isn't you?
A case of mixed identity,
you say?
A mutant, mad monstrosity
Wears the mask of poetry.
I see . . . and I believe,
For I have aged and wits have dulled
Since we last talked like this.
Yet you seem younger now than then
And even stronger now than when,
Seduced, I chose to live with prose
In cryptic caverns, musty holes
of scholarly pretense.

Too many tedious years I bowed
To pompous fictions of my proud
And fatuous philosophies
Before I finally confessed
They'd serve us best
If they were dressed
In jingling bells and waggling fool's cap,
Honeyed rhyme and jangling rhythm,

Cheap verse -
Not somber academic robes,
And moldy scent of august tomes
Declaimed in sumptuous, solemn tones.

Then far more years I bent my knee
Before the highest vanity -
Theatrics of theology -
Where, anxious and bedeviled, we
pretend to see
and strive in vain
the name explain,
That prophets dared not speak or spell
But barely gasped -
Jahweh!
Hushed whisper,
Startled rasp -
Jahweh!
Holiness ineffable!
How did I hope
We might define, confine, explain
Infinity in language, logic -
reason's chain?

* * *

O, Poetry!
Those worried decades made me old;
Six thousand years have made you young.
And I am flattered that you thought
It right to call me home again.
It's pure delight to meet like this,
Reintroduced by mutual friends,
And talk awhile to trace the past
And hope and plan a better future.

Poetry! Old friend!
Musing confidant of childhood.

THE MORALIZER

Behind the sneer
His eyes showed fear
Then blazed in anger.
Voice mounted pitch to ride
The singsong of derision.
"Man, y're just another one of them
Perpechul motion moralizers.
Everything - from sex 'n' family life
To power politics or culture -
You reduce ta right 'n' wrong.
Your first concern is always ethics,
tied ta every thought 'r feelin' .
'S'always morals - praise 'n' laurels
Or your scowl of disapproval.
You spout a constant commentary
On the world's affairs.
You're outa' sync, ol' man.
According to them old medieval
Ethical opinions that your
Value system still dispenses,
Everything I do or think must
Carry moral consequences."

* * *

The old man's eyes replied
In sadness spiced with laughter
While he said, "I tried.
I'll be more clear hereafter.

But, first, I thank you, friend.
You're kind and I'm obliged,
For you've been very frank.

I don't have many friends
Or foes who are so candid.
And with the state of culture now,
Such honesty is needed.
Till you spoke up, I didn't know
How well I had succeeded."

A CULTURE LOOSE IN WEIGHTLESS SPACE

This culture yaws in weightless outer space.
"We're free!" we boast,
and drift on unrestrained
And helpless, with no orbit, with no place
That offers to contain us.

We have escaped!
We're unencumbered - free
from Hebrew law
and old Greek logic -
Troublesome prerequisites
of sanity and truth.

We slither from the grasp of gravity,
levitate in fatal fantasy,
License manifest insanity,
in the guise of creativity.
"Ahh, genius!" we name
our maudlin megalomania.

We boast sweet liberation
from
the tedious dull facts
of human limitation
And swoon in sweet delusions that

        unfettered freedom can
Escape from every form or mass
                or presence greater than
                        our self-inflated self.
Thus egocentric emptiness
Sets out without the bulk or press
                of truth-
        Unencumbered, hence
                adrift in weightless space,
                locked and bound in void,
                a pointless, vacant place
        Where souls
                hang
                cap-
                tive
                to
                their
        free-
                        dom.

And, thus endowed, we've launched the proud
        autonomous individual,
        anesthetized amnesiac,
                into human history.

Skewered at the navel by
        his self-made center of gravity,
He coils and thrusts his useless limbs
        in endless blank vacuity;
Out where no other force
        resists his course, so he,
        insists, of course, he's free.

        Indeed, he's free -
                to flail and twist,

to draw and spread
his loose but useless limbs
around his inner axis -
One . . . isolated . . . one
denies the creeping fear,
boasting, "I am free!"
pleading, "Look at me!
or I may disappear."
He answers to no other mass,
no greater gravity, to hold and mold
and grant mobility
to whip-lashed whims and fantasy.

He's captive to himself set free -
or loose,
adrift . . . and disconnected -
One . . . abandoned . . . one
in weightless, empty space.

No place to set his foot
Or rest his mind where he might find
some small resistance that could free him,
He flails and flings his limbs and mind,
locked in helpless, specious freedom.

ONE FINAL INSULT

I heard one final insult cut and fray
The bonds of reason and morality.
Then something broke.
I heard the snap;
I felt the lurch.
I watched the gap
That split the church.
I smelled fresh hate

Infect the state.
But I was cool,
No pious fool.
So no one spoke
To guard civility and decency,
Or gird the values of the family.

And since that day we've all inherited
Precisely what our silence merited.

PRESUMPTION

"Well, if there is some,
Ahh, some . . . deity,
Then it or she or he must be,
We all agree,
Pure love and mercy.
And all this groveling piety,
This scrupulous anxiety
Is obsolete.
Passe!" you say,
And hope to let the matter drop.

But, let's suppose that mercy's way,
Despite our human lust for justice,
Will grant our greedy souls by grace
Fulfillment of each earnest wish.
What if. . .
You get quite all . . . you ever . . . wanted,
To find. . .
You never wanted more than this -
Your own . . . private . . . bliss?

Then as angels and archangels and all the company
Of heaven laud and magnify his holy name,

And hand in hand, the galaxies cavort
In orbits clasped in gravity's support,
While all the billowed, blazing spheres
Convulse in reverential tears -
The cosmos bows toward
Dimensions that afford
This universe a place
Within the firm embrace
Of other thoughts than space,
and time,
and matter.
Then cosmic intellects ignite
In tongues of fire and words of light,
And all the hosts of heaven and earth
Must bow to cry:
"Holy, holy, holy, is the Lord of Hosts;
The whole earth is full of his glory!"
"Holy, holy, holy, Lord God Almighty,
Which was and is and is to come!"

. . . And yet . . . you get
Quite all you ever wanted -

Your own. . . . private . . . blissssss.

* * * * *

Ineffable Holiness, eternal One,
Confront those who stand outside
The traditions of worship and scorn them.
Chasten those who stand inside
Those traditions and trivialize them.
Deliver this culture of modernity
From impudent impiety.
Deliver the community of faith
From concocted commotion

As well as dead ceremony,
From superstitious sentiment
As well as smug intellection.
God, demolish all our affectations,
And grant us sacred times when we
Can only tremble speechlessly
In awe before your holiness,
Knowing that more
Would surely be less. Amen.

IMMACULATE CONCEPTION

I see the comic's eyes prepare to dance.
"Immaculate conception! Ha!"

It's always good for jokes.
A smirk, a wink - quick sneer; don't think.
Derision is the jokester's game.
He needn't understand the name;
Pious words are all the same.

Next he'll sniff and grin, "Ha, ha, we're in;
You're out, if you don't know how smart
It is to mock the 'or-tho-dox.'
(O, that word! Hee, hee.)
Who still believe - Ha, ha -
In virgin birth - O, me!"

And patiently, the Church endures
The ignorance of arrogance,
The sadly malformed cleverness
That doesn't know the difference
Between the one and utter other,
Virgin born - the Son,
Conceived immaculate - the mother.

Not every virgin's holy, and
Not every saint's a virgin.
But such distinctions lie beyond
The self-deluded cleverness
Of those who nurse derision
To hide their indecision.

JOHN PAUL II IN THE AMERICAS

"Viva! Papa!
Hurrah! Hurrah!"

Some shout to shine
with borrowed fame.
Some come and bow.
Still more refrain,
resist, resent, and, last, recoil
from the honest words
and boldest claim
of this man's moral authority:

"The goal, my child, is holiness!"

* * *

His long reputed gift for language
gropes and fails to keep apace.
His trembling hands and shuffling gait,
faltering speech and rigid face,
Are tokens of old wounds and sorrows -
hard yesterdays and few tomorrows -
the scourging of prolonged disease
and grim responsibilities.

But sad eyes still speak one thing boldly:

"The goal, my child, is holiness!"

* * *

Next morning's news will fly the flag
of those who squirm and spit, and wag
Their swollen heads in chronic pain
of self-indulgence and complain,
with petulance and pout:
"It isn't fair! Your edicts would deny
the free expression of our *needs*. And why?
Whatever's done between (or . . . ahh . . . among)
consenting, free adults can not be wrong."

In sadness for the human race,
He bows his head to set his face
against this current currency
of trendy, waffling sophistry
And let his life speak yet more boldly:

"The goal, my child, is holiness."

* * *

But, we resist, for we have learned
enlightenment, modernity,
and moral cleverness
with chic irreverence
for ancient pious things.
And we insist,
"The sacred *id* must never be suppressed;
Our whims are *needs* and *need* to be expressed.
And chastity -
Ridiculous!
How feeble, old, and unenlightened!"
(O, we seem bold when we are frightened.)

But still his sad eyes weep
and penetrate

a nation's soul
with heaven's intent:

"The goal, my child, is holiness!"

THE SUBTLE, SEDUCTIVE ALLURE

What is this subtle, seductive allure -
This hazy hint of high romance
in smoky, dark cafes
behind the tinkling glass
and woozy laughter?

What is this smooth veneer of punditry,
This sticky smear of sophistry
at squalid psychedelic parties,
where impersonated genius
rides turbulent, tempestuous waves,
Or hushed and smoothed beneath
slow, heavy eyelids drifts
Limp and loose on wide, black lakes
of private fantasy?

What is this stubborn predilection -
youthful, willful, fierce conviction
that mystic wisdom, grand adventure
and cryptic, occult power lies
behind the glazed, benighted eyes
of self-destructive ignorance
that masquerades as wisdom?

Why do I acquiesce and listen to
This blase blather, drunk on self-importance?
Why is it so convincing to the hungry,
eager ears of devotees
who grant entranced attention to

this shallow self-styled oracle.
He staggers on the precipice
of psychic, dazed disaster,
commandeering followers
to prop him up and push him on
along the edge of suicide.

This new hypocrisy
purports to stand beyond
morality and reason
and thus beyond dispute.
This boaster puffs and flaunts his bold pretense
To knowledge of a better way
than common decency,
civility and law.
And he persuades because he seems
impervious to loss,
and names that "courage" - to confuse
his votaries who fail to see
That one who's lost integrity
has nothing left to lose.

INSULTED BY CHEAP PITY

A lazy, shuffling moral senescence,
a patronizing pity,
masquerading as compassion,
Excused his sin, refused his penitence -
the natural guilt
a healthy mind would own.

Too long he failed to recognize
that he had been insulted by
The sated sentimental song:
"But you have been deprived of our

opportunity to learn
The niceties of middle class
comportment.
You're clearly not responsible
as we, of course, would be."

Too long he acquiesced, until
some grit met steel within his soul
where ire found the fire to say:
"Enough! I'd rather be a man,
willfully depraved, not pitiably deprived.
I choose to be accountable.
I claim my shame and own my own
responsibility.
Don't ask me to consent
to pity that excuses me
to keep me pitiable - obligatory victim.
I choose to be a man,
Not your poor, flat caricature.
I will to be intentional,
Accountable - authentic self.
I will not trade that self away
for your sweet condescending slush.
I refuse to be excused -
and robbed of moral manhood.
I am guilty!"

THE BONSAI MAN

The bonsai man,
his meager life
meticulously
dwarfed,
then pierced and pruned by custom's knife,
accommodates what's vogue,
what's modish, trendy, chic.
He keeps what's hip today,
but cuts and burns last week.
And, strange enough, somehow, it's deemed as "in"
right now to look
neurotic and disheveled,
partly crazed and all bedeviled.
So bonsai man does what he can; he bows to
stunting style and nods
to tortured, twisted bonsai fashion,
wincing at the common clods
who misconstrue his eager passion
to play the role of wizened soul,
refusing nature's ample ration
of truth and grace, as free as rain,
and fertile nutrients that rise
through human roots in common soil -
community, the warm surprise
of plain civility and toil,
and gratitude in neighbor's eyes.
But bonsai man despises norms of happiness and
health,
And faking shriveled, wasted forms, takes
sympathy by stealth.
So, dwarfed and pruned he joins the popular
complicity
To live in quaint caricature of authenticity.

LOST IN BOSTON

Briefly unattended on
A public counter, they were gone,
Stolen, we supposed, because
They looked the part of someone's
trendy status symbol -
"cool shades."

They had indeed been costly lens,
But, in fact, *too* dear
For use by any but the one for whom
They were prescribed.
To help her failing eyes appreciate
A long awaited tour
Of old historic Boston Town.

* * * *

The irony of theft, remains
Inherent in the deed.
The thief? We thought we might propose
A profile that could lead
Us, not to that lost item, but
To those lost truths that seem
To have escaped a culture that's
Misplaced its self-esteem.

So somewhere, we conjectured, was
A stumbling sneak whose eyes were strained
And skewed and out of focus,
Who, struggling to look smart, looked pained,
Unscrewed and grandly bogus,
From peering at a fuzzy world
Through stolen misfit trifocals.

A spectacle in stolen spectacles
        Should stand out in the crowd, we thought,
Until we noticed all the other souls,
        All red eyed, bleary, and distraught,
Who strut and strain to fill
        Strange artificial roles
Because they cannot see
                        Their own intrinsic goals
        Through stolen, misfit attitudes.

A CHILD WAS BORN

A child was born today
        In that city we call "The Angels."
Not fully born - but torn away
        Then tossed away
To be borne away
        With the refuse of . . . Los Angeles.
O, God, what's in a name?

So he died and never cried,
        Though he whimpered once for shame,
        That those who called him
                        into being
        Never called him by a name.

And he bore a strange regret
        That they could feel no shame,
        Those who called him
                        into being,
        Just to be . . . without a name.
O, God, what's in a name?

****

Another child was born
                On a distant, different day -

A day still named for him,
In a place called *Bethlehem* -
"House of Bread" or "House of Fighting"?
That, we still debate and must decide.
O, God, what's in a name?

Announced and hailed by angels -
That child, how he was named!
Savior, Christ the Lord,
Rose of Sharon, Lilly of the Valley,
Bright and Morning Star,
Fairest of Ten Thousand,
The Word, Good Shepherd,
Light of the World.
The Way, the Truth, and Life.

Named and nurtured,
Loved and worshiped
By those who knew his names,

Then cursed and slandered,
Scourged and tortured,
By those who dared not know -

He bowed his head beneath words that read
"Jesus of Nazareth, King of the Jews."
O, God, what's in a name?

Thus, he was also torn away
And borne away, like refuse
from Jerusalem,
But he remained so rich in names
That we who come this late in time
Still take his name, a paradigm
Of all we hope and pray and claim.
O, teach me, God, what's in that name?

* * * *

I yearn to know,
Because that other child,
        Nameless refuse of Los Angeles,
                Was found today -
Part born, then torn., still warm
His final, whimpered breaths
        Were heard and heeded by two sisters,
        Poor Clares of St. Francis,
        Angelitas en verdad de Los Angeles.

They swaddled him in cloister's robes,
        Then touched his head with tears and rain
        To give his mournful life a name.

And thus the refuse of Los Angeles was borne
                away by angels
And welcomed in the name of One called
        Wonderful and Counselor,
        Prince of Peace, The Mighty God,
        Everlasting Father.
        King of Kings and Lord of Lords.

O, God, my God, what's in a name?

LATE TO THE FLAPPER'S FUNERAL

The hushed bustle of condolences
        Had withdrawn to the chapel.
The "remains". . . remained -
        Across a room the length of eternity,
                As vast as space-time,
                Light years removed -
When an old friend hustled to the door,
Then halted, eyes pleading - searching the emptiness,

The awful chasm between . . .
Between extremities he could not name.

He had hoped to be early,
And here there should have been
Little islands of gossip
And warm, congenial shoulders -
Safe, sandy beaches with broad smiling harbors
As he navigated this dreadful distance.

Breezy congeniality
Should have filled his tattered sails here.
But he was
Arrested instead, frozen in dread,
Stalled in the strait twixt decision and fate.
He clutched the doorjamb,
Suspended in time
By a thread of eternity,
And peered across that empty immensity,
Hailing the casket with an anxious stare
That gaped like that vast vacant room
Now stretched between them.

* * *

Born to the hope of a new century,
When the past was haunted
By phantoms of authority,
They were young and choosing their way
In the "roaring twenties."
And their way was "hip,"
So clever together,
In those liberated years,
"Enlightened" years,
The moonshine years,
The jivey years, the flapper years,
O, God, too . . . many . . . years -

Drunk on the hubris of modernity:
Fantasies of smartness,
And obsolescence of wisdom,
Anility of all tradition.

He came now, half a century later,
To this dreaded meeting,
Soul and body bearing still
His vast agglomerated
Conventions of the unconventional -
Still indentured to some pseudoliberation,
Still in bondage to defiance of tradition.

Oily white hair, slicked back to the twenties,
Made a long gesture to the sixties,
With a wave up front to the fifties
And a duck tail salute behind.
Taps on his shoes,
Intended as a bold self-announcement,
Served here to announce
Each fidget of fear.
Trousers spanned tight,
But flared cuffs fluttered
Below loose, tremulous knees.

* * *

He clutched the doorway, shocked -
Stare fixed, fixed, fixed
On that dreadful . . . dead-ful box.
Then, at last, his thin frame flinched with another memory.
Another time - there had been death ahead
And shame behind.
And memory called upon some power
Of military discipline:
"NOW!" it cried, "Move out! Go! Go! Go!"

Now! . . . But how? How? How?

One foot ahead, he paused again
Until his fingers took the lead
And snapped a clever syncopation,
Answered by his taps.
"Look a' me, man.
I'm cool! I jive!
Ain't scared a' nothin'."
A jazzy little dance step
Moved him through the door,
Tap - one, two. Snap - three, four.
And . . . across . . . the . . . floor. . .
Tap - one, two. Snap - three, four,
Tap - one, two. Snap - three, four.

He jiggled . . .
The length of that awful chasm
Between jive and reality,
Then slumped . . . and gaped,
Suspended in grisly ambivalence
Between horror and sorrow
Where, there within reach of firm, cold truth,
All cleverness failed.
The syncopation ceased.
The awful tick of real time commenced.
"Stay. Stand. Don't cave, now.
Stay a decent time, man.
One - ten thousand, two - ten thousand,
Three . . . four . . . five.
Just . . . won't . . . jive."

Sharp, quick seconds snapped and tapped
Their terrible tribute to time until,

Eyes clamped - tight, taut, and fiercely dry -
He denied a sob,
But expelled a breath
That choked like death,
And fled again from hard reality.
Tap - one, two. Snap - three, four.
Tap - one, two. Snap - three, four.

THE COLLAGE

Museum piece, so smugly ugly -
Oddly, here preserved.
This dusty gray and musty brown collage
Offends the eye and mind because . . .
It could have meaning, even beauty
Left alone
And spread about by chance,
Or hanging from the creaking attic rafters
High in some old well-worn home,
The residence of recollections.
There, draped with authenticity,
Memories, and love,
Arranged in random lines of warm
Voluptuous nostalgia,
Shaped by serendipity
And covered with the dust of years,
It could have granted mystic credence
To an honest photograph.

But posing here "so frightfully refined,"
This rag has been resigned to dirt - confined
To vain presumptions of the modern mind -
Pompous striving to exert
Some show of non-conformity,
Pious efforts to achieve

Irreverence.
Arty sentiments assert
Themselves too much,
Endeavoring to look
Profoundly artless,
And managing to be all too self-conscious -
A speck of useless spectacle
Here jammed into the public eye.

THE LEADER

The product of our schools of higher yearning -
Like pining for celebrity, affected eccentricity,
Ostentatious novelty as ersatz creativity,
Committed existentially to ardent amorality -
He was a man uniquely qualified
To rise and stand . . . obliquely to the tide
Of popular opinion,
A masquerading minion.

He faced somewhat upstream, but yet toward
The distant shore for which he feigned to ford
The currents of this age
Devoid of heritage.

In shallows, crouched to simulate deep feeling,
He bobbed, neck deep, half standing and half kneeling.
Appearing thus to stride
Toward the other side,
He drifted.

STATE OF THE UNION, 1999

(On the State of the Union Message during the Clinton impeachment hearings.)

The state of this Union, nineteen ninety-nine,
Regrets to imitate
The galling, baffling state
Of this man's mind -
Brilliant, privileged, needy,
Wealth of opportunity,
Poverty of soul,
Inveterate madness,
Consummate sadness of man and nation -

Sadness for one so remarkably gifted to govern
Who can not govern himself;
Sadness for this one distinctively affluent nation,
Still destitute of self-control -
Hence, self-esteem;
Sadness for morality -
Tangled tight in covenant with self-esteem,
Braided in interdependence,
Each cause and each effect
Of each the other
And staggering as they lean into a slippery future.

The state of the Union, nineteen ninety-nine,
Mirrors this man's mirror mind -
Splendid counterfeit,
Counterfeit splendor,
Moral bathos glamorized,
Moral impotence disguised
As robust, virile pride,
"Weakness . . . being lied
into something meritorious." *

This culture makes a virtue of its weakness,
Pompous pageantry of moral bleakness.

Political apotheosis!
    One man's hubris has seduced us.
Now we watch
    His craving need to be desired,
        Urgency to be admired.
    Misfortune of success -
    This climber reached too high.
Bad jokes are not forgot
    In that clear upper atmosphere
    Where he stands proud - and fatuous.

He charmed his way to dominance,
Clamored up this prominence
    Where now he stands exposed,
    Above the cloud that would gladly shroud
        A humbler man from history's glare.

Exposed! Fate grants his fondest wish -
    A national joke,
    Buffoon in baroque;
        And history waits
            With probing, prodding patience.

In the state of this Union,
We're embarrassed to hear
    That strong voice raised -
        Assertive, clear -
        Blaring the ring of high purposes.
And we're diminished to see
    That strong hand raised
        In warranty,
        Wearing the ring of sly promises.

Splendid and bright, gifted and cogent,
Lustrous potential . . . smeared with disgust,
    Chronic indulgence,
    Prolonged adolescence
Spells well the fate of man and state,
A state seduced by moral arrogance,
    Not sensuality.

Gulled by moral pedantry,
We buy the lie that sexuality
    Should suffer no constraints
        For such confirmed elite as we,
        The natural aristocracy,
    Modernity's enlightened ones.

Such fantasies of liberation
    Haunt our blighted generation,
Conceived in self-deceit
And swollen with conceit.
    We quash the moral memory
        And decent sensibility
        That grace the common masses.

We strut our limping disability,
    Prance our individuality,
    Parade our artsy eccentricity,
While proudly swaggering bored contempt
        For those who merely walk
            Through life
        And neither want nor need
        Our special moral license.

Thus here we watch
    Our leader illustrate
    The baffled, bleeding shape

Of this poor staggering state,
Exemplified by elitist pride
In moral impotence.

*Friedrich Nietzsche, Spoken of the moralists in *Genealogy of Morals*

A SOMETIME RHYME

Repeatedly
I've heard the boast that ours is not
"A rhyme time."
Presumably
The gravitas of modern thought,
(Like prime time
Television) disallows
Such artificial gimmickry -
As rhyme and rhythm.

But sometime rhyme and full-time cadence
Well up in our speech,
Driven by some deep
Reserve of mind beneath
Our consciousness.
And they will block the natural pace of thought
As much when we deny the place they ought
To hold
As when we give abject obedience
To worn-out forms for mere expedience.

AMISHMAN

Greetings to you, Amishman,
Young Amishman,
Detained like me, both waiting for
A late M.D. who's snarled in noxious,
Blaring city traffic.
We find our mutuality
And smile to note the irony
Of what they'll call "emergency"
In modern medicine.

But while we're both impatient patients,
Why do I observe another
Apprehension on your face,
Trapped here in this too public place?

O, my friend,
Don't look with envy
At this culture's trash -
The garish cheap parade,
Inanity, insanity -
Blaring, glaring in the flicker
Of the omnipresent tube
That molds the morals
Of the nations.

Don't wonder at the pseudo-tough
And pseudo-smart
Pretensions of "the Ain-glish"
Crowded here around you.
But pray for us instead.
We're chained to affectation's lies
And hate the few who dare surmise
We're actors, fools and hypocrites,
Wannabe sophisticates.
Even here, surround by

Reminders of mortality,
We still deny reality.

Amishman, young Amishman,
Don't gawk with envy or pretense like us.
And don't avert your eyes to act as though
You didn't know
Or did not look.
Please, don't destroy my sweet illusions of
Your once magnificent renunciation.
Don't smudge the luster of your liberation,
Coveting our cheap-thrill titillation.

O, keep your ways -
For us!
Esteem your days
For us!
And cherish the embrace
Of regal limitations.
Clasped in family
And faith's community,
You have a place to plant your feet
and stand and stretch and be
A man amidst this groveling madness of modernity.

Ach, mein Freund,
Stand firm and straight and strong,
And ride with self-esteem,
A pride that's justified
By honest, simpler ways
Than our disordered days.
Clog the frantic traffic of our highways
And our minds
To slow our pace and make us question where
We need to go
That horse-drawn buggy cannot take us there,
And just on time.

II.
A LITTLE ANTIGRAVITY

"Angels can fly because they can take themselves lightly."

G. K. Chesterton
Orthodoxy

SHEER SOPHISTICATION

I seethed as city traffic crawled.
I scourged the traffic light that stalled
        My lane, but then
Revised my attitude a trifle when
An apparition, slithered into view
        Within the mirror set above my eyes
                And strictly angled for the view behind.
        I glanced and glimpsed a fantasy that lies
                Behind the conscious mirror of my mind.

She came adorned in lustrous, burnished black:
        Her sable hair, her sheer chiffon and chic
Accoutrements - the glistening Cadillac,
        The jeweled necklace - pearls on jet and onyx.
                O, she
        Was pure personified sophistication.
        So svelte, so au courant, so utterly
                chichi.

With just the right hauteur she smiled, I thought,
                At me.
I gaped, entranced.
        But better thoughts and higher obligations
Persuaded me; chagrined, I quickly glanced
        Away a thought or two,
Till one last glimpse, it seemed,
        Would be a fit conclusion.
But no! O, sheer misfortune!
        Cruel disillusion!
I'd turned in time to see that peerless face
        Proceed to evanesce and disappear,
As, quickly, coalescing in its place,

There rose a huge pink luminescent sphere. . .
Of bubble gum.

That loathsome stuff! The world's deranged!
That troublesome
Pink bubble gum
Was just the first
Of bubbles burst
Before the traffic light had changed.

SHORTFALLINGS

He was a mild young clergyman, in fact, a youth
At heart, and unprepared to tell the painful truth,

Too meek to speak of *sin,* preferring to resort
To euphemism, cooing how "we all fall short."

Just short of pure perfection - who could be offended?
But, O, how artfully our self-deceit's amended!

I've heard of one gray Sunday morning - fated day -
When he, with high decorum, grandly rose to pray,

"O, God, we've fallen short." And then, with humble air and
Thinking on his feet to form the plural gerund

(I must believe it was by providence of sorts)
He chose to ask, "Dear Lord, forgive our falling shorts."

THE SOUND OF CRASHING SYMBOLS

Before I learned to read, I learned to hear
    And see and feel the stories in
        The echoes of the conch shell by my ear.
    It was "the ocean's roar."
So I was told, and it was easy to believe
The varicolored spirals of the shell still held
Some faint reverberation of the stormy sea,
And echoes of those rolling, roaring depths, still welled
Up in that symbol in my hands.

And listening, I could visualize
The distant heaving ocean's rise
    And fall, the cumbrous cryptic dance
    Of water's mystic high romance.

* * * *

My sons inherited the fantasy.
    My father held the shell to grandsons' ears
    While I recited through nostalgic tears,
"Now listen, boy, and you can hear the ocean."

Thus my sons were taught
    Without the benefit of science -
        Aural or acoustical
    Analysis of echoes and
        Distortions that resound in small,
            Enclosed and empty places.
    None of that!
Symbolic meanings led the way,
    And I complied, convinced this myth
        Was still superior to many
        More they'd learn about what's called
                reality.

My sons should know the folklore first,
    No thought for dull pedestrian facts.

* * * *

And thus the matter stood
    One hurried Sunday morning when
    The clock had failed and I had urged:
"Look, sleepyheads, it's very late.
Cooperate. Don't hesitate.
    Wake up.
        Get dressed.
            Be prompt.
                Don't squabble.
                    Don't delay.
    Hurry is the order of the day."

Then ten or twenty minutes passed,
    And all went well, I thought, until,
From where I listened on the stairs,
    The children's room seemed much too still.

But just before I would have called
To reprimand, I was forestalled
    When one small dawdler's cries
    Announced with keen surprise
And wakeful tones, long overdue,
    "Hey, y'know, you guys?
I hear the ocean - in my shoe!"

&

He was an ample ampersand
And self-assured enough to stand
    Where he'd been dropped at top of page,
    Orphaned at a tender age -
Alone, detached and understated,
Unemployed and unrelated
    To the busy crowd below.

But other characters complained.
    It irritated them to see him stand
    There, single, bulging slightly, head in hand,
    A solitary, all unsightly. . . *AND. . .*,
The punctuation marks explained:
    "An ampersand must understand
        That we have expectations.
    He shouldn't lead, but still he'll need
        Sufficient following.
    The best conjunction tries to function
        Subtly in the middle;
    Or pushed up front, he still should want
        To exercise connections."

Next, dollar sign opined and whined,
        Narcissus' own high priest:
    "With just a little flair and flourish,
        A curly tail, at least,
    Why, you could be a treble clef,
        A musical artiste!
So sing your song or move along -
    Shape up, you ugly beast."

But Ampersand could not be scared,
And Ampersand would not be wheedled.
Neither threats not blandishments
    Induced him to affiliate,
    And finally it was too late.

Offended by his impudence
    A mob of letters organized
The Group for Citizens' Defense,
    The G.C.D.,
    To oversee
Such things as number, case and tense.
Then just to deal with Ampersand,
They made a federal offense
Of loitering disguised as *AND.*

They rigged a jury of his peers -
Like dittos and quotation marks -
    To firmly guarantee
    Quick unanimity,
    No pondering or pauses.
Their verdict called for forty years
    Of labor chained between
    Two independent clauses.

Though Ampersand at last
Could understand his past
    And grasp the sheer enormity
    Of hate for non-conformity,
He still refused repentance.
    He quietly suppressed his rage,
    Defected to another page,
And never served his sentence.

NUCLEAR WORD GAMES

Nuclear. Impressive word!
It has the ring of power. And I prefer to hear
it ring, ring loud and clear - *clee-ar.* I am not
reassured when
those who proffer high credentials, representing
agencies, committees
and commissions of international import, pro-
nounce the word to sound like
jo-*cular*
verna-*cular*
instead
of *clear* and simple
nu-*cle*-*ar*.

I WATCHED AN OLD MAN

I watched an old man's hesitant approach
　　From far across the room,
And, watching, understood his compromise
　　With wincing, aching joints,
His brief debate with pain and gasped replies
To each reluctant lurch that pushed the pain
Ahead just far enough to meet again
　　At every other step
　　Toward the distant goal
Of one man's lonely indoor marathon.
　　With eager empathy
I watched that old man flinch, then angling nearer,
Reach to lean on me behind the mirror.

GREAT VOLLEYBALL

I gulp and gasp.
Chest pounds, resounds.
My throat and temples throb.
With muscles tense and eyes alert
I judge the ball for speed and angle,
Twist to miss and let it pass,
Then turn to watch. Good choice.
"Outside!"

Our serve is fast but hard returned.
My abdomen contracts:
I crouch to lunge
Then stretch to slide,
And friction burns
The ribs I ride,
One arm extended out in front
To place a hand
Between the ball and floor.
I feel the impact, watch the bounce
And hope a teammate's leap
Will clear the rolling body underfoot.
I plant my hands;
A twisting push-up instantly
Returns me to a crouch
Where thighs snap tight,
Knees flex then speed me
Back into the game
As someone shouts, "Our point!"

Then, peering through the net,
I hear the thud of solid serve behind.
The ball is fast returned
And volleyed anxiously -
Over once, then twice, and more.
Hips twist; the tendons
Of my knees and heels snap tight like wire
Then flex

To drop me back again
Into a crouch, prepared to lunge
Or charge the net.

My left foot planted,
Stable to the side,
Then turns to point the way,
With right toe poised behind.
Knees flex then drive toward the net
Where quivering thighs delay an instant,
Judge the time, and then contract for launch
As calves produce the final lift.
Toes leave the floor.
I'm high; I fly.
My hammer fist and wrist descend
with power.
I feel the spike
Propel a plunging ball.
I watch it strike the floor
And ricochet clear out of court!

We win again!
I cheer. I soar.
Three games of four.
"OOOOooooEEE! Great volleyball!"

* * * *

But then I pause to catch my breath.
I sit to stretch sore muscles and
Accommodate old joints.
I dialogue awhile with pain,
Then struggle to my feet again,
Where, finally
Descending from my bleacher seat,
I hobble to the floor,
Congratulate my grandson's team
and, last,
Tell him how well he played
Through every move I made.

III.
PENN'S SYLVANIA

"And Nature, the old nurse, took
    The child upon her knee,
Saying: 'Here is a story-book
    Thy Father has written for thee.'"

H. W. Longfellow,
Fiftieth Birthday of Agassiz

SILHOUETTE ON FIRE

Against a sky of fire I saw
One solitary winter oak -
A charred black, cinder silhouette
On blazing gold and violet -
The sentry post of one bleak raven.

Yet sharper than the raven's caw,
That barren, charcoal profile spoke
Of gnarled beauty's black regret
Slashed through winter's late sunset -
Darkness, deep in splendor graven,
To make that luminous bold fire
To ache with numinous desire.

THE COOL, CLEAN CRISP

The cool, clean crisp
Of days like this
Portends the hour
When autumn's power
Slips slowly into winter's death.

Yet, bold to hold their grip on life,
The humble mums and marigolds still sing
Their dazzling brilliance,
While tall gray trees can only drone
a trace
of clinging color
that scampers down
their sheltered side
and nether branches.

But fallen reds and yellows
      Mingle deep in heaps
            Along the fence row,
Where flinging streaks in flying, dancing, swirling leaps,
      Of wild intent, show
            hues of
                  loyalty
                        to whirling wind
                              that brought them low,
      To dull and die, then green again,
      Through earthy modes of modest brown.

* * *

Then, beaming bright, broad faced approval
      Of the scene, a brilliant blaze
      Of sunlight's late luxuriance
Emerges through low clouds
      That now have lingered countless days -
            A mercy cloaked in darkness.
      For colors as they were last week,
            In this illumination
            Would burn the mind
            And turn this kind
                  Of day to conflagration -
            Beauty's fire and frenzy.

DEEP SILENCES

Wrapped within the brook banks curving arms,
Caressed by waves of swamp grass,
Draped in drooping pine boughs -
I am accepted and immersed within
This world of wildness - yet a spy.

A spotted fawn,
Just old enough
And bold enough
To venture out of sight
From mother's care,
With instincts wiser, though, than I,
Returns my gaze without reply.
But ears that rise and turn,
Invite me too to learn
Through his bright sensibilities,
And I comply.

I smell the fragrant, fresh wet moss
And feel the cool of whispering breeze.
The water whimpers wild pathos,
Caressing rocks and nursing trees.
I watch and feel and listen,
And, listening, learn like fawns
To love

The silences, deep silences, within
The melodies -

Times when pileated drummers cease,
And cardinal's careening call
Desists
With heckling crows' and blue jays' rasped
Interrogations.

And, last, the distant warblers pause,
High overhead.
On nature's cue,
The concert stops
To listen to . . .

The silences, deep silences, within
The breath of wild-song.

And in that tranquil pause I hear
My father's whispered wood lore
Echo down the brook banks of my memory:
"Listen! In the woods we always listen."

Memories gather, drifting deep.
Nostalgic stillness calls; I weep
Hot mystic tears and wait,
And watch - and listen to

The silences, deep silences, within
Nostalgia's yearning.

* * * *

But now the fawn has listened first
And heard before I could
The ugly threat and boasting blare
That violates
This tranquil air.
Ears rise and turn, eyes search until
He firmly plants one small front foot,
His instincts sounding danger,
Then bounds just twice to disappear,
Thus leaving me, the last to hear
The dread cacophony of fear -
From beasts who dare not venture here
Without some crass distraction from

The silences, deep silences, within
Their emptiness.

They come here, indolent and loud,
Spinning wheels and blaring proud
Abusive noise - crude techno-toys,
Amplifying
Strident steel and rasping sax,
Perpetual percussion.

Backfire! Motors fulminate -
JOSTLING,
BOUNCING,
SPINNING,
POUNCING,
CRASHING,
SMASHING
Down the very path . . .
that
fawn
had
tread
so
ten -
der-
ly . . . .
They come on crashing, crushing wheels
That depredate and desecrate the songs
Of life they never hear
Who crush their own tormented minds to stop
The silences they fear -

The silences, deep silences within
The song of truth.

* * * *

The fawn is gone; the noise is passing,
As it will.
But yet the woods reverberate a
Lifeless chill,
While I
Still hold the shattered silence to my ear,
And hide my blushing face to weep - and fear
For what I am . . .
And what my race has thus become
So soon since we were weak and wild
And wise as any nature's child,
Keeping silences, deep silences, within
Our Eden.

MIDSUMMER NIGHT'S SCREAM

The silence of midsummer night -
A cricket punctuated hush
Rolls and stretches, gentle, quiet -
Soft warm belly up
Toward the moon.
Night-silence is a sleepy kitten,
Rampantly relaxed,
In dreams of warm security.

A tranquil purr, the chirp and buzz of life,
Extends across the fields and deepening woods
Beyond my window - open to
Nocturnal wild, nature's child,
Soft warm vulnerability.
Stille Nacht.

* * *

But silence bursts!

The hush is ruptured by
One final, pleading, bleeding cry
That cuts the sleepy black of night,
Spilling bright primordial fright.

Terror's sharp vibrato,
Serrated edge of scream,
Death, a brief stiletto,
Slashes through my dream
And lacerates the dark.
In one small voice, relentless horror
Cuts and twists through pulsing fear -
Severing the artery
Of pity.

My spirit bleeds, convulsed
In throbbing hemorrhage
Of impotent compassion.
Viscera ache, vertebrae quake.
Ribcage resonates
Acoustic sympathy,
In psychic empathy
For one small dying screamer in the night.

Respect the anonymity.
Young rabbit taken by a cat?
Or kitten taken by an owl?
It will remain
Unknown, unnamed -
A faceless sacrifice . . .
That represents us all.

For I hear, echoing behind
That anguished cry, still undefined,
A universal wail:

Gasp of captive prey . . .
or starving predator,
Sob of battered child . . .
or guilty batterer,
Groan of dying soldier . . .
in counterpoint with foes.

* * *

O, God, how can I pray?
I fall in bafflement,
Stricken by this lethal
Silence of the night.
I do not choose to kneel
But stumble to my knees,
Where, hushed, I must observe
Mute deference before
That wailing, wandering chorus,
Plodding through the strife,
Vagabonds and pilgrims
Exiting this life
In final unity -
Community of suffering.

For even while I watch,
That caravan of sorrows
Treads its somber way
Back into the silence -
Submissive to reality,
Receptive to a sanctity
That intimates eternal home
In mercy's ultimate shalom.

Dumbfounded by anguish, astonished by grace -
I wait in mute submission to
The plea behind each dying breath.

While those who suffer speak my prayer,
O, God, help me accept my death
In this dark mystery of silence -
Silence of midsummer night.

AUTUMN'S FALL

The glow of color, smothered now,
The burning days of autumn bow
To cat-tail swamp's dead beige and brown,
While naked oak's gray tangled frown
Droops on bleached-out brier thickets.

A huge blunt weather front of gloom
Slogs slowly up Ohio Valley,
Settling down to make a home
With me.
We'll watch and wait for winter sun
To angle under pregnant snow clouds,
Stretching noonday shadows long
Like gnarled fingers clutching northward.

But while we wait as though
Some shadow's shape on snow
Could point us on our way,
We acquiesce
To brief bleak days, long murky nights,
Dark minor key, spare melody -
The plodding dance of melancholy.

WHITETAIL

The snow is soft and deep,
Muffling ruffled leaves
And brittle twigs' complaint
Beneath my feet.

I steal a silent step, then pause
And wait.
I must be closer to my prey
To gauge his rack
And take good aim.

My prey! A healthy, whitetail buck
Investigates the soil.
He digs and nuzzles deep
To reach a squirrel's cache of acorns,
Garnished with chilled wintergreen
And wrapped in juicy roots.

It's only while his head is down
And busy with the task,
I dare to risk a step - then halt
As still as stone and hold my breath
When wide eyes rise and seek to find
The source of some stray scent or subtle
Sound I don't detect.

At last the dark gray antlers show
Against the winter light of snow,
But he is taut, alert:
His great eyes search for me.
I fear he hears my quickened heart;
I know he'll see a burst of breath
If I can not restrain
My hungering lungs.

He stomps and snorts.
White puffs of caution rise
Above his regal head.

And I must contemplate his robust lungs
And warm resilient heart
That I . . . intend . . . to stop.
I question my presumption:
What is this human right I claim
To turn this vibrant thrill of being
Into . . . meat?

But oddly now he's reassured,
Content again to satisfy
His hunger.
He slowly drops his head
In one last bow to me -
While I,
With careful aim, reply;
And we
Both shudder, simultaneously,
He in sudden death and I
With dark responsibility.

WINTER DAWN

Winter morning, diffident dawning!

Horizon glimmer, frosted flame,
Reluctant simmer speaks of day,
    Sings of light; and I concur,
For I believe what mornings say
        of hope and probability.

But schedules push me on;
I haven't time to listen to
    the song of dawn
    and color's rhyme.

Head drops, breath quickens;
I converse with snow.
    The frigid whimper underfoot
        Soon chirps as I march double time
        Toward that steep predestined climb.

I lunge ahead, white frozen puffs
    Preceding in the chill
        And leading to the hill
            That steers my steps - until . . .

        Secluded down a darkened lane,
        Deep in valley's lengthened night,
        A genial yellow kitchen light,
        Gentle glimmer, warms, cajoles,
        Excites the mellow brown bouquet
        Of coffee perking sluggish souls
            To color groggy gray
            With wakened shades of day.
I pause . . . ; I know I'm welcome there.
    Imagination sniffs the air

And tastes the hot alluring brew,
But rising moans of bitter wind
Impel me on toward a dawn
That can not rise complete until
I lift my eyes atop that hill
And see from there again
Where I too long have been
And where, please, where I'm going.

I'VE KNOWN THIS DAY TOO WELL

I've known this sharp-edged winter day too well,
Too intimately.
I've walked too close
His frigid side.
Our bony elbows clash
Like dignity on pride.
His blades cut deep and paralyze.
His frozen whispers whip my eyes,
Compelling steamy tears
That fall . . . and freeze . . . still promising
To nurture trees with warmth of spring.

I've stalked these tall tree paths,
And watched this frozen lakeside dawn to dusk,
Ear by ear with caterwauling wind
And creaking, ice-bound oaks.
I've seen the spruce and fir with upstart steeples
Point and hail the high procession,
Clouds' brisk, disciplined advance,
Like gathering armies, hosts of warring peoples -
Then watched that solemn, columned march of snow
Descend into a swarming mob below.

* * * *

But now, for one brief moment,
Brilliant sun rides high
On deep, broad incandescent blue,
To reconcile high puffy whites
With driven, driving blacks and grays -
Fierce, colliding
Intertwining
Lower clouds
Of ebony inlaid
With ivory, like braid.
I gape into that mobile face of change -
The rush to cover blue with white
And white with cresting waves of gray
And black that tumble mighty walls
Of heaving white to earth again,
Where I retreat beneath
These strong broad hemlock arms
To watch
With reverence,
And something very near
To memory and fear
For those who traveled here
A few brief centuries ago -
Those more intimate with snow
And storm, who had no easy flight
Before the killing cold of night.

* * * *

Then, once again, the passing wave of gloom
And snarl of wind and snow
Withdraw before
An angled spear
Of light
That strikes
The lake -

One dazzling splash of white
reflected,
Dancing colors - light
refracted.
A truce! The wind and storm are hushed.
But I still watch suspiciously.
Waiting first to listen,
I hear before I see
Fresh troops advancing audibly,
Among the tall trees west of me.
Assembling next above my head,
And hovering for reconnaissance,
This storm observes from tree tops first,
Until, in one brief furious burst,
It dumps a shuddering weight of white -
and fearful beauty -
Over every form in sight.
And then . . . it's done
In less than one
Short hour.

* * * *

Astonishing transfiguration!
Beauty's power to redeem
Confounds a world resigned to face
The lash of wind
Through dismal days
Of gloomy grays -
Relentless winter.
Incredulous,
I stare in awe at what I saw
Transpire before my eyes.
This fresh voluptuous world of contoured snow
Softens every shrill protest below.

While, instantly, the overcast above
Retreats before a brilliant white battalion -
Cotton cumuli on blue.

The world's reborn,
And I emerge
From under sagging evergreens
To meet and touch and taste the white
And greet resplendent, dazzling light
That shimmers on the face of heaven
Smiling on the transformation
One brief storm can bring.

THE CHARACTER OF OAK

Given space and sunlit grace,
Oak is peaceful, broad and round -
No will to rise and dominate.
His arms reach wide; his shade is full.
With rugged, armored hardwood trunk
He feels secure,
And frightened creatures
Seek his shade and shelter.
"Come in," he says, "you're welcome here,"
And brawny branches clasp the fear
Within a twisted, tight embrace,
Contorted elbows, nesting space.

With ample sun, oak keeps his place,
Stocky, strong and satisfied.

But given hostile competition -
Hurried hemlocks, pushy pines,
Anxious aspens, clinging vines -
Oak's knobby arms will stretch and reach,

His gnarled fingers clutch for sky,
Protracted trunk climb vertical and tall.
And his reward? Our selfish praise:
"Good lumber! Long beamed hardwood timber!"
For certain he'll be felled and sawed,
Then dried and planed, reworked and fitted,
Smoothed and stained so we might see
And read his coarse-grained history
In broad, assertive bands of growth
That tell the epic tale of oak.

IV.
LAMENTATIONS AND HOPE

"Love's not Time's fool, though rosy lips and cheeks
Within his bending sickle's compass come;
Love alters not with his brief hours and weeks,
But bears it out even to the edge of doom.
    If this be error and upon me proved,
    I never writ, nor no man ever loved."

William Shakespeare
Sonnet 116

FOUR STRINGS ACROSS THE CHASM

So many years, as child and youth,
My son invested in
This timeworn violin -
A family heirloom,
Useless, waiting,
Still - until it found in him
Someone to love and master it.

I touch and feel the memory
Of mellow, anguished beauty,
Ecstasy, vitality - such power
He wrenched from this slight, fragile thing!
Reminiscence gropes and probes the past
For lingering melodies.
I brush the still, taut strings.
The full, arched belly sings
A throbbing, sweet nostalgia, deep,
Reverberating tones of . . . grief.

So many empty, aching years
Have passed since that son died.
And I have never dared till now
To hold this fragile instrument
So close - for fear it would hold me,
Clasped too tight in sorrow's might.

But now, this delicate old elegance
In clumsy hands,
I strive to be deliberate,
To think objectively,
Mechanic's instincts first.
Release the strings; it seems so rational.
Relieve the tension captured there.

Uncertain how, first, gently I
Caress the aged wood
And gradually recall,
As timid, untaught fingers
Close upon a peg
And imitate what I have seen him do.
A minute twist, a gentle tug, it's free.

"Hnnnungh! What? But . . . what?" I gasp.

I'm startled by this greeting, this unexpected meeting
With the tug of thin, tight string, that frail small thing
That spans the awful chasm set between us.
I'm startled by the touch of him I grieve,
As fingertips are brushed by energy
That *he* placed there.
Potential, waiting, through the years.
Life force bearing still
his fingerprints.
Long, strong artist's fingers reach across the gulf
Between my life in space and time
And his . . . beyond those trifles.

The force of his life presses back against my grip,
Resists my pressure, once again;
Exerts his independence, once again;
Affirms his strong, fierce ardor, once again.
How well I recognize that touch, that firm
Familiar force,
That vibrant source . . . of death-defying love.

Yet closer than these thin,
Taut strings - I hold within
My hidden mind and memory,
Still stored in deep and secret strife,

Dark energy in minor-key,
Keen radiants of his bright brief life
That haunt my mind and search my soul,
Reaching, seeking melody,
Reaching, seeking harmony,
Reaching, seeking pitch and tempo,
Startling with remembrance,
Clasping me in reverence.

So much of his intensity
Is wound too tightly, deep in me.
I strive to loose the tensions there
And find obscure dimensions where
our past and present meet.

Today I grapple with his art and will and skill,
Just as I wrestled with his proud
young body then.
And I must box his nimble mind and spirit still,
Just as I sparred against his quick,
strong hands before.

And, once again, today, as always,
I'm knocked down -
O, by mistake, of course,
just as he always said.
But, once again, here on my knees,
stunned, on the ground,
I'm pained but proud of that son's strength
and genius.

YOU TAUGHT US WELL

You taught us well, dear friend,
these fourteen months,
to look toward our end
with grace and dignity.
Letting go old ways,
You charted, graphed and illustrated
fresh new authenticity.
It's been a long and arduous course
in spiritual maturity.

You taught us how
To cry with love
and laugh at pride,
To ponder human destiny
and own our own mortality.
You showed "The Way,"
and tolerated no
dishonesty - no cheats -
No artificial highs of eager sophistry,
No self-adorning lies of cheap enthusiasm.

You taught us well, dear friend;
We knew our lessons thoroughly
when you were through with your long course
of suffering
and died.

RITES OF PASSAGE

A twisted world where parents bury children
Reverses nature's order, runs awry
    Of everything that's just and right.
And thus I wept for one ill-fated son
    Who slipped away in coma's night:
"But *he* was meant to be there when *I* die."

Sad years since then I've wept the words and fought
The tears in silence, grappling with the thought:
That son was meant to be there when I die.

But as I bow to age, and that time nears,
I've learned to quench my rage and dry my tears,
And, chastened, let a deeper, wiser cry -
The certitude of love - recite it still:
That son *was* meant to be there when I die,
And now, by grace, I dare believe he will.

WIFE

I balk, I choke, I gasp
    In mute finality,
Convulsed by every hint
    Of her mortality.

All queenly, fierce and gentle,
    She strode into my life,
Part lioness, part fawn,
    But every part . . . a wife!

We've sung the hymns of hope,
    And wept in disbelief
We've laughed on peaks of mirth
    And sobbed through depths of grief.

We've climbed on bloodied feet
    Up brutal shards of stone,
But never since our vows
    Has either walked alone.

I balk, I choke, I gasp . . .
    At that tormenting thought.

* * * *

Together - we have traced
    And mapped the crooked paths
        Of grief and death and dying.
        Clutching those who died
          With those who grieved, we cried . . .
Together . . . and embraced
    The spasms of their fears,
    Co-mingling tears with tears.
Together - how we clenched
    Our lives together through

The tempest of our grief!
Together . . . half a century
Together now.

But truth will tell when one must grieve alone.
I balk at that.
My proud commitment to
Reality collapses, founders.
I war against those outlaw thoughts,
Vagrant desperado phantoms -
Ineffable inevitable.
Yet I am ambushed in that narrow pass
Where all must travel
Single file.

God, teach me to be spare with words
Of confidence and faith -
True words, but only right
The other side of pain.
God, help me to submit to real
Compassion - suffering with -
And let the long, strong, cross-drawn arms
Of Christ embrace us all.

GRIEF

Bleak, black grief
Claws and rips the shredded flesh of life
And love from bloody bone
And mires my torn remains in murky strife -
Disconsolate, alone.
I sink in hopelessness; I ache and agonize.

But spare me, please, my friends, your guilt-averted eyes
And wistful pity's gushed attempt to neutralize

The grizzly truth with gracious well-intentioned lies.
I loathe such counterfeit condolences
Solicitude contrives
To cover bleeding wretchedness with balmy
Sentiment that strives
To mask the heavy, caustic stench of doom's
Decay with oleaginous perfume's
Sweet reek and unctuous artificial blooms.
I retch against that thick narcotic haze
Of smothering incense
And choke through mawkish, suffocating days.
My strangling soul dissents:
God, grant me one cool gasp - brisk winter's breath
Of Truth - to brace my mind to face this death.

I want, I will, I choose to bow my head
In stark eclipse of artificial light
And vow to suffer truth - an honest grief,
A tangled, tortured tribute to the cost
Of love . . . and excellence of what I've lost.

I will hold fast to grave reality,
Not levitate on flimsy fantasy,
But prostrate, forehead to the stone,
Stretched cruciform in gaunt submission,
Face this pain, embrace this death,
And yield to truth's stern admonition.

For though this grief is agony,
I will not dodge reality.
I've bet my life on him who spoke
Into the teeth of his own death:
"I *am* the truth."

V.
NOSTALGIA OF AGE

"The leaves of memory seemed to make
A mournful rustling in the dark."

H. W. Longfellow,
The Fire of Driftwood

THE DOLL

My father sat, sad eyed, confused, each breath
Ensnared in veils of dim uncertain
Memories, the tattered curtain
Shrouding those dark years before his death,
While we who loved him had so soon become
A nameless blur of trusted faces,
An aimless purr of gentle voices,
Pushing hard against the edge of night.

That day three sons and father sat to search
Behind, beyond each other's eyes -
As though to hold and memorize
That homespun moment's essence lest it lurch
Next overboard to sink
In his abysmal sea
Of black forgetfulness
That held too soon, too much
Of family history,
Long lost like grains of sodden sand
Once squeezed too tightly, held too close.

Then, suddenly, my little niece danced by,
Indifferent to our brooding melancholy.
All our aching memories,
Were likewise lost on her -
That bright young life
All facing forward,
Heedless of the past.
She skipped straight through the looming gloom
That filled our corner of the room,
Then, just as soon, marched back,
Now followed by a tattered doll

She dragged along by one torn foot -
To her, a slight uncertain thing
To hold and hug - or twirl and swing.
But when she turned away
And capered through the door,
The dolly fell and lay,
Akimbo on the floor,
And in an instant impulse
Father roused from somnolence
And moved with anxious haste.
Deep inherent empathies
Had thrown him down on reverent knees,
Where, scooping up the drooping thing
And, cooing baby talk,
He comforted - the doll. . . .
And I, held helplessly aside,
Ensnared in my surprise
Beheld a blaze of bright compassion
Burn the haze of dreary, ashen
Stupor from his eyes -
Those eyes now suddenly aroused
By deep paternal instincts drawn
To hold and comfort . . .
Me!

Answers loped ahead of questions!
Eagerly, I wept the thought -
How I, the helpless first-born
Fragment of his life,
Once was placed into those arms -
A gift from that proud wife
Who watched again now from another room
And saw once more the gentle, fervent light
Of fatherhood illuminate his night.

For him, for that brief, tender moment, I
And that small tattered doll were one,
New life, new love, fresh image of his flesh,
And proud responsibility.
That doll was I, and I was glad to rest
There still where strong, lean arms had lifted me
And he had laughed and cooed his love
Unnumbered times to dry my tears
And comfort childish hurts and fears.

I wondered if he might not hear my cries
As in my soul I wept:
"He can't look up today and recognize,
But I'll rest there, still kept
Secure in his proud love, though broken
Like that doll - my name unspoken
Since nightfall.

"I pray, O God, embrace us all
And please permit me to remain
There by his heart and dare again
To love him in his helpless time
As much as he loved me in mine."

HILL TOWN, RIVER TOWN, HOME TOWN

Hill town, river town, Pennsylvania home town -
River clean and narrow here,
turning,
twisting,
ambling
down -
early on its journey to
the great Ohio,
mighty Miss'
and briny Gulf.

Hill town, river town, little hustler settled down,
first on flats where creek and river met,
challenging the flood plain
with proud downtown pretensions.

But then she saw the heights
and asked, "Why not?"
And climbing up on stilts
she occupied the slippery oily
hills with rambling, rawboned derricks,
followed by rough sheds and shanties.

Conquering first the gentle slopes,
she coveted the heights.
The hills were ambushed
once with winding paths,
Then charged straight on with steps
And roads too steep for wheels,
and finally outflanked and occupied
by devious trails - whereby
I come back home today.

* * * *

Stealthily, I come at dawn,
Gaping voyeur,
scrutinizing
White fog fingers of the night
That cling and linger into light,
Still fondling proud green mountain breasts,
Spilling milky billows down
To cover shabby gray and brown -
The faded shades of yesterday's
downtown.

Five floor buildings once stood proud
beneath the slopes,
all occupied with dignified
bright offices to congregate
above their thriving stores and banks.
Now upper floors are empty but for broken hopes
of yesterday's raw wounded pride
and brokers of cheap real estate
with little cause to render thanks

To this rough little hill town, river town, home town.

* * * *

I make a quick retreat
along a pockmarked street
of broken brick, now
shedding scabs of asphalt.
This town grew old like those of us
once born here.
Intending to run straight,
we found that
nature's obstacles
and human willfulness
soon led through
tortured twists, precipitous trails

where whimsy mists conceal the rails
and bridges that could bind
and integrate the mind.
down on a humbler plane.

But winding down these streets
a little torque is good
to tease and, some may hope, delay
intruders from the flat-lands:
checkerboard cities,
straight, flat streets
with numbered names and boredom.

Hill town, river town,
still proud of old run down
schools and hospital,
still loyal to your churches.
Your schools still train another generation -
to abandon you;
Your hospital still greets the newborn,
heals the old, or, failing that,
provides another exit;
Your churches, new and old,
clapboard store fronts,
crumbling stone cathedrals -
Exhort, reform, and point toward another way.

And, though, the places where a city kneels
to worship, also bow to passing time,
Still, faith assures and hope allures,
for here they know life comes to go,
and here they celebrate the there,
a glistening, golden city where -
Beyond another river, and
Upon another hill -
Community endures.

RIDING BACKWARDS UP THE RHINE

Moving back through space - and time -
Riding backwards up the Rhine,
    I first arrive in Germany
Astride the rhythmic click of wheels on steel -
    Hypnomagnetic track
    That draws my spirit back,
While I'm compelled to face and look
Toward the way ancestors took
Two hundred fifty years ago,
    Those who packed away my genes
    And rode these selfsame river scenes
    On down to find their way across
        Atlantic storms
        Looking forward,
        Watching shoreward,
    With an ancient code of courage
    And an eager love of labor.

Today I'm back where they began,
    And I am now the alien,
    A guest in Rhineland villages
Where family roots and family name survive.
But, though a stranger here,
I feel and love. . . and fear . . .
    Familiar spirits - sentiments that rose
    Within these hills and resonate with those
    That tremble in the marrow of my bones,
    To call their wandering, wondering children home.
And though this pride of place and family
Are chastened now by painful history,
    Some old allure remains
    As warm as mother's milk

Or blood.
It marches with the thump and throb
Of tuba's um-pah-pah
Or soars with searing melodies,
The beauty and the brawn
Of Bach or Beethoven.
It resonates in Luther's thunder,
Sobs through Spener's piety.
It gasps at Goethe's raptured fire
And Eckhart's mystic paradox.

From *"Ehrfurcht vor dem Leben"*[1]
To *"Deutschland über alles,"*[2]
Devout or bellicose,
The souls of honest *Volk*[3] are found
Immersed in piety or drowned
In boisterous beer.
From cold, clipped, chipped precision tools
Of icy academic rules
To wisdom's fire and ecstasy,
From science to philosophy,
Theology to poetry,
Some sad, fierce pride in *Vaterland*[4]
Still calls my name.

Despite that bent black slash of fear and pain
On memory - the warped, embarrassed stain
Of Swastika that flew above these bright,
Green villages, lush grassy hills, the white
Peaked Alps and regal Rhine -
Here still resound
Old words still bound
To racial memory's profoundest passion -
Mutter. Vater.

Blut und Boden!
Ach! Ich liebe dich. [5]

* * * *

[1] *Reverence for life*
[2] *Germany over all.*
[3] *People or folk.*
[4] *Fatherland*
[5] *Mother. Father.*
Blood and soil!
O! I love you.

KISS ME

"Kiss me; I washed my face," she said,
    And raised a wispy, weary head
To search her daughter's gentle eyes
    For sanction - or some compromise.
Large eyes, broad smiles, their features seemed
    To interchange as nature schemed
To help them comprehend each other -
    One the child and one the mother.
But which was which? Old roles reversed,
    And soon new lines were well rehearsed.
So lifting mother's wrinkled brow,
    The daughter plead, "Each morning, now,
Please scrub your teeth and wash your face.
    Be organized, and don't misplace
Your slippers or your upper plate.
    Yes, dress for meals, and don't be late . . .

And don't just grin at my tirades!
    Where have you put those hearing aids?"

The mother's eyes reflected - there
    Beneath a wreath of faded hair,
The features of her only child
    Whose great eyes teared and soft mouth smiled,

"Kiss me, I washed my face, you know."

I SEE NOW

I see now what I've never seen before -
Before this gauzy ghost of gray crept in
To bend and blur the clear sharp edge of things
I hold at center, focused - sight and mind.

I watch a grisly pock-marked face appear
Each time I blink and stare at snow or search
An empty page for possibilities.
Those bulbous features briefly coalesce,
Stare back and read the future in my eyes,
Reflected there where future lies,
Constrained by hard new boundaries.

And even while
My mind interrogates this waiting page
For thoughts to weigh and words to visualize,
That face stares back and sees a shadow age
When I may read no more than phantom eyes -
Abysmal gray opaque,
The shade of burning thirst.

I see now what I've never seen before -
And, first, I see new urgency
To read and write
Before my light
recedes into
Black premonitions of a stumbling time,
A humbling, bumbling time
When long beloved and fond familiar features,
Bright young faces that delight
My hopes, will bend and fade for me
And droop in sadness I can't see,
For nothing sparkles then
like memory.

But now I see beyond the fear and sighs,
    And see because that phantom face
Crept in and frowned so I would clear my eyes
    To look beyond, toward a grace
        I could not even guess before
    Where snow-bright winter solstice laughs
        At bleeding autumn's woeful lore.

And when I look beyond and see this way,
    Perceiving in each loss some gain,
I must believe that death itself will hold
    A grace sufficient for its pain,
For I have met that angel here, already,
    Secreted in every little
    Death, like this gray gauzy ghost
Secluded now inside the once bright
    Sanctuary of my eyes.

I HAVE TRAVELED ON A FRAGRANCE

I have traveled on a fragrance,
That wispy thing of phantom wings
That soars above the years and miles,
Past long forgotten tears and smiles
Of friends who walked with me
On rain-wet paths of fresh-cut hay,
Down fragrant trails through evergreens
Or blooming mountain laurel.
Landscapes and old portraits from
The past wait there within
That dusky gallery
Of vintage memory
Where fragrance is the key.

The scent of new sawn wood or turpentine,
Sweet spice or slight bouquet of gentle wine
Can move my mind
Across the years and miles more readily
Than anything I feel or hear or see.
Olfactory magic!
Neuron powered flight,
Sparked by faint aroma
And fueled by nostalgia.

O, I have traveled on a fragrance.
Years and miles whisked by, and I
Returned to times and places that
I knew from childhood's quaint perspective -
Eye level to a table-top,
Knee up to mount a chair
head first.
And there I was a seeker of
Hidden places, open spaces,

Always
under foot or over head,
behind piano, under bed,
high in trees, on roof of shed,
at top of sliding board
or winding attic stair -
Dusty, dry scents there:
the sun-baked books
andbundled letters,
doughboy's uniform,
faded flags and
yellowed photographs.

And then, when tired of rambling, I
Became a seeker of laps
And patient, gentle arms
That held me into sleep
Then carried me to bed
Where fragrances of blossomed farms
Like perfumed mist invited me:
"Come, learn to ride the balmy waves
of childhood memory;
Return and live this time and place
again . . . again . . . again."

How often since,
I've traveled on a fragrance
Through the wood lot there
And past those hills and valleys,
Following one small trout stream east,
Then rippling down a creek to join
A forest river's flow
That laps at all my early years
And lingering recollections.

Some aromatic essence of
      Residual memory
            Lifts and carries me
      Along her perfumed banks
      To places and to times
That taught me who I was . . .
      And who I ought to be.

There scented throngs still greet me.
A terraced bank of roses
      Cheers my late return.
And tall, proud palace guards,
      Sentinel hollyhocks in
      Stiff, straight rows welcome me
                    back home -
      Here to be
            and live again
            what I have been and lived before:
Catcher of honey bees
      And mighty, buzzing bumblers,
The limit I dared to squeeze
      Into old glass tumblers
      Or canning jars with makeshift lids
Where pollen shedders, stout old scolds,
      Hummed alarm and anger
Until I granted them reprieve
      And fled their droning danger.

O, I have traveled on a fragrance -
      Over rolling hills and decades,
      Lifetimes, oceans, continents -
To times more vivid now than then,
More sweetly incensed now than when
      They touched my childish wonder with

Their aromatic bitter-sweet,
And perfumed memories -
Redolent nostalgia.

I REMEMBER POVERTY

O, I remember Poverty,
his diabolical intent
to stunt and maim the innocent
and terminate the joy
of childhood's first prerogative -
Security.
He lurched and grabbed to steal my youth,
But, firmly held at bay
By parent's skill and happenstance,
He skulked across the way
Where I observed his countenance,
Hovering by my dearest friends,
his shadow darkening strong bright minds,
young incandescent souls I loved.

And in their father's hollow face
I watched a strong man's fear and shame
Withdraw, reluctant to embrace
Those lives that clutched his grim, gaunt frame,
To hide from ghosts of bleak disgrace
that haunted him.

Yes, I remember Poverty.

And by remembrance still today
I touch the tears, and feel the blood
of blistered hands and bony arms
draped deep in dust of wornout farms.

Though dust and sweat have turned to mud
And clog the back roads of my mind,
I still remember Poverty.
I see him tear the smiles away,
and claw the shivering flesh of dread
From raw, red eyes and quivering lips,
still whispering secrets in my head.

While further down that valley town,
I smell the odor of despair
in tattered clothes and matted hair
and lives of patched-up disrepair.
I hear the curse, the cough, the cries,
See love's deep melancholy eyes
down alleyways and valley lanes,
through shattered doors and window panes,
Where once I watched bleak Poverty,
besmeared with shame and grief and grime
in one small town for one brief time,
When I was still too young
to fear.

ROOSTER'S CROW

The strutting rooster crows the surest
Self-esteem and shows the purest
Pride I've seen. O, I admire
His confidence and cocky fire
For mate and territory.

But how I've wished his puny brain
Could comprehend the joy and pain
Of sharp nostalgia that he brings
To mind for me each time he sings
Of hens and chicken coops.

No other voice can quite cajole
My mind like his, transport my soul
To those small villages and farms
That hold me still in memory's charms
And melodies of childhood.

SHARP EDGE OF HISTORY

I see my trophy briefly flash
its angled edge among the rounded, softer stones,
A forearm's length beneath the face
of mountain water, dancing mirrors.

I reach and clutch, but silt, aroused
from sleep, leaps up to hide its cache.
My grasping hand drives blind and deep
within the biting rubble,
too eager for
the chiseled edge
of arrow head.

That edge retorts. I wince
to see the water red.

My anger riles. My tongue reviles
the crude, inanimate thing I grip
with blood and mud and malice.
And how should I have known,
And how expect
this lingering malevolence,
proclivity for violence?
Even splintered flint should mellow,
buried here for centuries,
then washed out into light
by recent summer storms.

My wound protests;
My pain interrogates:
Who fashioned this serrated edge
and shaped it to an arrow head?
Who bound it to a feathered shaft.
Who drew it hard against the spring
of yew or hickory?
Who poised that power, then released
and lodged it deep
In severed flesh
And splintered bone?

And whose, the flesh and bone?
Was it white tail venison?
Was it rival tribesman?
Or did that missile
Serve to slow the bloody flow,
Atlantic tide of clumsy pride -
the pallid bearded ones,
My forebears, who intruded here

and meddled - just as I
have done today?

* * *

Interrogation stalls;
Stoic stone deflects my probing stare,
Scorns my questions,
leaves me puzzling
to surmise
Some homicidal history,
petrified audacity,
that glares behind this chiseled face.

I still protest.
My overtures did not deserve
that sharp reply.
I reached in peace, to rescue -
lift a bit of broken past
from muddy murk
beneath the gleam
of mountain stream.

I simply hoped
to offer this
mean tempered chip
a place in human
memory.
But rancor lurking deep in old hostilities
preferred seclusion,
told me sharply
not to trespass.
Crude, rasp-edged killer struck again
And cut the hand that hoped to help.

It should have lost that sharp, aggressive mood.

It should have smoothed, caressed by centuries.

* * *

So I still ask,
    Who shaped this tooth-edged point?
And why and who, I want to know,
    Who felt its driving bite?
    Who clutched his wound
        and touched warm blood
    Then heard faint burbling death's replies
        to life's dread ultimate surprise?

I wonder; were there answers then
    when unembellished men
        lived close this rocky soil?

For I have been aloof too long,
    too blithely comforted by lies -
        the fraudulent security
        of opulent modernity.
The arrow dropped, my bleeding stopped,
        a sterile bandage hides my wound.
Detachment binds the hemorrhage
        of eager curiosity.
I watch indifference nimbly turn
    the tourniquet on hot concern,
    and, once again, I will not learn.

A GOOD MAN

(a ballad)*

1.
"Old Doyle, he was a go-od man,
A good man in his day.
Yep, Doyle sure was a go-od man,"
I heard his old friends say.

We sat there by his casket; and
I answered as I should,
"Yeah, Daddy was a good man," but
They all misunderstood.

They'd seen him tame the bullies, watched
Him hunt and shoot and box.
"Yeah, quicker than a wi-ldcat,
An' smarter than a fox."

But I had known Doyle better, and
I knew the gentle side,
And how he hurt with those he loved -
Like the day I almost died.

****

REFRAIN:
A workin' man, a fighter, Doyle
Was agile, lean, and fast,
But, oh, he proved a better man
When all of that was past.

And those who knew him best were blessed
    To watch him growin' old,
When fists of steel and iron grip
    Let go and showed the gold.
****

2.
"That Doyle, he was a re-al man,
    A real man in his day.
Yeah, Doyle sure was a re-al man,"
    I heard his old friends say.

We sat there by his coffin, and
    I answered best I could.
"Yeah, Daddy was a real man," but
    They still misunderstood.

They'd seen him dig and climb and build;
    They'd watched him work with steel.
But I had watched him strive and grieve
    Till all he was was real.

Yeah, Daddy was a re-al man,
    No boast and no pretend,
He was nothing artificial, just
    Authentic to the end.
****

REPEAT REFRAIN
****

*The melody can vary greatly; but, the first three times the word *man* appears in each verse (in the colloquial use of "good man" or "real man") it helps to interpret the irony of the lyrics if the chords progress from the major tonic to a minor based on the sixth.

VI
IRONIES OF MINISTRY

"The Christian ministry is the worst of all trades, but the best of all professions."

John Newton

BORNED AG'IN

A gnarled and withered life,
The old man was deprived
and ignorant, I thought.
I watched his naive trust,
abject credulity,
Walk forward, tears assenting to
that arrogant evangelist
whose pride could never see
The simple power of honesty
Incarnate in that one old man
who knelt
and wept in silence.

With pride I thought I saw straight through
the lies of stage performing preachers.
But this old man's rough artless soul
saw more, for he
saw clear beyond
The sham that strutted there on stage -
and clear beyond
The swaggering, boasting insolence,
Lathering, blathering, here to hence.
He saw beyond
The clever entertainer's words
to One who is Eternal Word,
and Truth incarnate.

* * * *

By what rare gift do simple seekers grasp
The substance and the truth of Christ concealed
behind the boasts and pranks
of prideful, stage-struck preachers?

I'm puzzled, but I can not doubt
they do, for I can show you still,
in one small, squalid shanty down
beside a 'Hio River town,
One simple, honest man who will
recite his story,
"hyare ta glory,"
as he says,
"An' halleluuu, an' bless the Lord.
'Cuz I 'uz changed and saved f'um sin
at 'at 'er meetin' - borned agin.
An' ever sinct I never bin
the same,
'Cuz Jesus die f'r me,
'n I gon' live f'r him."

And, stunned, I bow,
for I know how
he lives and strives,
And in his honest, bumbling words
I sense more wisdom than I've heard
in all the tightly reasoned speculations
that science or theology
contrives.

THE CONFESSION

She blushes and shifts her soft shoulders,
Black tresses caressing white breasts.
Her flaunted confession still smolders.
Her eyes make forbidden requests.

She weeps over sweet stolen pleasures.
She weeps, but she does not repent.
I halt, for I have other treasures.
I halt . . . but this thought won't relent.

* * *

O, God, forgive!
How smoooothly fantasy
Slips into the featherbed of desire -
The deathbed of truth!
How simply I could rationalize
This one indulgence -
So meaning*ful* I must . . .
So meaning*less* I may.
Lies go either way,
And the Father of Lies
Winks and leers
At a distance.
While I grapple up close
With the Angel of Truth.
God, I welcome the wounds Truth inflicts,
But please grant me his blessing also. Amen.

CONTINENTAL DIVIDE

Above the lush, low continent
Of human lust and pride
And stretched across its full extent,
There stands a great divide -
But not between the innocent
And guilty - each on side.
For those who glimpse their own intent
Must know that they have seen
All human will's predicament,
Yet mountains rise between
The few
Who hold them selves accountable
And those
Who make excuses.

WEDDING CAKE

Conjugal legalities
are lawyers' stock-in-trade, not mine,
and not the church's first concern.
What Mother Church proffers is grace,
the power to live the truth and learn
the depths of human interface.
May God forbid she try to be
the peddler of her holy rites
to all polite society
Or universally require
false homage to her holy fire
by pagans of modernity.

She ought devoutly to refuse
to marry those who won't recuse
themselves,

although they're impotent
to elevate a bashful wish
much less an honest prayer's intent
to build a life and home upon
those sacred vows they wish to make
for sentimental fashion's sake.
Be sure, O, Church; do not mistake
Communion bread for wedding cake.

A BOLT OF SILENCE

A scorching bolt of silence strikes through to the core
of casual conversation
And leaves us fazed and dumb
before the whelming roar
of speechless consternation.
One stray and reckless word
you spoke and we both heard
with mutual surprise
Revived old memories
of long forgiven lies
and human frailties.

The thought has caught us unprepared.
We sit here witlessly ensnared
in blank, embarrassed hush.
One hollow word reverberates,
And nothing else communicates
above the thundering silence.

ALL I EVER WROTE

How many thousand heavy pages past
I toiled with might to write theology
And strove for truth and strictest clarity
Of reason and of diction.

But when I learned my subject was too vast
For words, I turned to write philosophy
Instead - and wrote so well 'twas clear to me
My true gift was for fiction.

But, three long novels past, I've learned at last
That all I ever wrote was poetry,
To call the dance of love and irony
In cadenced contradiction.

CLERGY BURN OUT

Good God, I'm weary -
Soul-spent, bone-bent weary,
But on Sunday I must be up-lift-ing.

From thirty days, with no day free,
No hour of rest from min-is-try,
Good God, I'm weary -
Soul-spent, bone-bent weary,
But on Sunday I must be com-fort-ing.

I'm weary -
Weary of calculation,
records and registration,
program coordination -
the grinding administration
of this volunteer organization . . . called church.

Who made me father to
these wrangling mid-life adolescents?
Who made me referee
in their games of vanity?

Good God, I'm weary -
Soul-spent, bone-bent weary . . .
And on Sunday I must be in-spir-ing.
Good God!

LORD, HELP ME LEARN

I brooded, angrily complaining while
I paced in raging haste to burn the bile
that roiled resentment's horde
within my soul.
I stormed, protesting, "Lord,
I don't deserve this ridicule, this vile
Contempt - the petty ignorance and lies
that slander and despise
your church, the saints and all
that's holy, just, and wise."

But there I paused, and pondering,
Heard those words reecho in my head
Until I slowly fathomed what I'd said
And prayed, "Good God, it's true; I am a fool,
And don't deserve this flattering ridicule.
Please help me, Lord, to earn
This spite I tried to spurn."

TRAPPED IN HIS OWN WORDS

Trapped in his own words! He knew
the instant that he heard himself
that he had spoken foolishly -
Too hastily, with words that had
been contemplated many times
though never granted sanction.

But suddenly, in anger,
there they were,
dancing in the sizzling air,
charged with old resentments.

I saw the consternation
sweep his face
and watched him contemplate
his options,
knowing he could stop there, where
he stood, to own his anger and
apologize.

But that, he thought, would leave him looking weak,
and he was never strong enough
to recognize his weakness,
uncamouflaged reality,
true self.

He showed just one brief glow of blush
and one short shade of fear
Before his vanity could hush
the truth and domineer
his mind and soul and conscience -
compelling them to follow
words that knew no reason.

HUMILIATION

I skimmed the page too quickly, read it wrong,
And, hearing lines and tune, still missed the song.
    I recognized the humor well enough
        But, laughing, failed to see
        The poignant irony;

Thus, misinformed, I made a judgment
    Based on partiality,
    Preference for stasis,
    Familiarity as ideology,
    My vested interest in . . . sheer ugliness.

I chose to see the ignorance and strife
    That I imagined from the start,
And I retorted, sure I had just cause,
    The moral right, to speak my mind.

And then I watched my caustic, proud replies
    Burn deep
Before I could apologize.

* * *

Have mercy, God, forgive my pride
And more, please grant the wisdom of
    Humility, not mere humiliation.
Embarrassment will not suffice
    To save me from that peevish pride
        And narrow self-distortion
        That grieved my friend
            And wrenched my soul
            Into this snarled contortion.
Enlarge my heart and mind
    With will to wait and understand
    And wit to welcome chastening.

I DID NOT BLEED

I did not bleed on Mount of Olives,
Did not feed the arid soil
    Where he sweat red submission to
    The nail and thorn of human scorn.

Where he dropped down in torment,
    I casually . . . dropped by . . .
    And dropped a camera -
Tourists' badge, contemptible
    Confession that I'd soon forget
        Without a glossy-faced portrayal
        To lodge the place in memory.

Casual tourist, I . . .
    Dropped by, where he dropped blood.
And then I dropped the camera -
    No more than that - and felt no more
        Than sharp annoyance, shattered glass,

        Till . . . last . . . some vagrant, wild regret.
            Insane! But yet . . .
It was the madness of a superficial tourist's
    Mystic discontent
        That those bright biting shards
            I gathered did not pierce me
    Deeper, leave my blood
Where I . . . dropped by,
    And dropped my box of memories.

I came, I saw, I fumbled;
    But no pain cut me deep enough
    To leave my blood in that parched dust,
            Land of Judah -

Land of martyrs, zealots' lust,
    Imperial invaders,
    Holy war crusaders,
Mutually branded infidels,
Hallowed hermit's torturous cells:
All have fed the appetite
    Of this dry soil for blood,
The thirst that greedily consumed
    The life I name my Lord.

VII.
CALL TO WORSHIP

"Thou reignest in glory; thou dwellest in light
Thine angels adore thee, all veiling their sight;
All laud we would render: O, help us to see
'Tis only the splendor of light hideth thee."
Walter Chalmers Smith,
The United Methodist Hymnal

HOLY, HOLY, HOLY

"Sanctus, Sanctus, Sanctus,
Dominus Deus Sabaoth . . . ."
I hear ten trillion whispers raise
That holy hush of urgent praise.
Mighty voices strive to rise,
Then melt away to tears'
Distilled ineffable replies,
As speechless love coheres
In tongues of flame
and rapture.

"Holy, holy, holy,
Lord God of Hosts..."
Incandescent intellects
Reply in timeless, spaceless grace
To worship One still stretched a-cross
That awful gulf between chaos
And ordered substance
On arms that cradle time and space
And matter in divine embrace.

Bound by love,
Those arms still stretch that he might grasp,
And then extend to heaven to raise
The cosmos in his hands and gasp:
"This is my body . . . , this . . . my blood."
And those who dare to hear still praise:
"Sanctus, Sanctus, Sanctus!"

HAIL MARY

Hail Mary -
In silent stone serenity
You welcome this poor city's poorest
through these lofty, crumbling arches
to the sacraments of life.

Full of grace -
You're wreathed in city soot and smog,
jarred and crushed by traffic's press,
besieged by hustlers, bustlers, thieves,
privy to their lechery,
acquainted with their blasphemy -
Confidante to truckers, cabbies,
street gangs, vagrants, prostitutes.

* * *

I flinch.
My spirit hurts to see
This tortured incongruity
Of pure maternal majesty,
Grimed with grim modernity -
Gritty inner-city squalor.
My soul revolts to hear
Our Lady's regal grace abused
By blaring pandemonium,
Her hallowed face insulted and
Assaulted by perpetual din.

But here she stays;
Where else to better speak
The power of purity?
For she's been here before,
Crushed and still triumphant.

* * *

Blessed art thou -
As blest you were to face the
Village gossips,
City throngs,
Vengeful priests,
Maddened mobs,
Brutal executioners.

They howled and raved, determined by
demonic fears,
The root of hate, to rend your heart
and mix your tears
With blood of him they loathed.

But you were blessed by
Seraph's song, celestial chorus:
"Hail to you who chose to be
'The handmaid of the Lord.'
Consenting, 'Let it be to me,
According to your word.' "

* * *

How just that you who were despised,
Ignored and scorned, unrecognized
By those who tore your heart,
Should be entreated still:

Pray for us sinners,
now and at the hour
of our death. Amen.

O, THERE IS ONE

O, there is One toward whom
by deepest nature's need
we're driven to rise,
But we mistake him if we visualize
vainglory in the skies
and name that "God,"
Or posture on the stage of cheap celebrity
and garish wealth,
As though we might expropriate integrity
by lies and stealth
and name that "honor,"
Or imitate and adulate the superheroes
comic books
Portray, while meek contorted conscience, moral
nescience, never looks
(and names that "courage")
Beyond cheap pretense to the fire of holiness
where souls aspire
To rise to him revealed on Calvary.
No star blazes higher
Than that divine prerogative.
Our birthright is to be Godlike
And comprehend that grace as *"godly."*
We're called to be like God indeed -
Not puerile myths we've fantasied,
But love's bold willingness to bleed.

COMMUNION OF SAINTS

Each week come meetin' time, prayer meetin' time,
    With tear-lined face and weathered hands she stood
    And asked, "Please pray for me. I want to be good."

No fawning here to smart society
That winks and flaunts its chic impiety.
    Just, "Pray for me. I want to be good."

And I, while still an awkward boy, I saw:
    A holy radiance in her candid tears,
    An honesty that braced hard, troubled years,
    An unaffected triumph over fears.
I saw and learned respect
        For piety and awe.

And now she's gone, long gone, and I am old.
But memory and sadness make me bold:
    I never said, I love you, friend. I should.
    And pray, oh, pray for me, I want to be good.

* * * * *

Lord, I am astonished by
    The fearless honesty
    Of that good woman's
        "Hunger and thirst for righteousness"
And equally astonished by
Her love's responsibility.
How many years she prayed for me!
I should have prayed for her.
I want to still, although she's gone.
    And what's this rigid Protestant reluctance?

How can I help but pray for loved ones
lost to death?
How can I think their intercessions cease?
Lord, I need not be theologically astute.
I need to ask my saintly friend,
"O, pray for me; I want to be good." Amen.

THIS MUSIC

Some music penetrates
Gray vacant space
With ordered time.
Some music permeates
Blank empty time
With ordered space.
Some brutal noise defies
All order.

But this . . .
O, this sheer luminescent surge of sound,
This mourning wail, this raptured sigh,
This rich voluptuous sonata bound
In covenant with love's reply,
Still undulates between the mind and brain,
Through body, soul and spirit - joy and pain -
Suffusing this brief time and space
With ambience of truth and grace
Eternal.

HYMNS OF WORSHIP

THE INCOMPARABLE

1. Before primordial energy,
   And after all shall cease,
Unchanging in the strife of time,
   Eternal Love grants peace.

Beneath the darkest ocean depths,
   Above the loftiest peaks,
The farthest, nearest - Cause of all,
   All-present Word still speaks.

2. Beyond remotest galaxies,
   Yet in the atom's sphere,
Resides the omnipresent One
   By whom all things cohere.

More feminine than woman's grace,
   More masculine than man,
The timeless, spaceless, formless One
   Attests eternal plan.

3. Our urgent, ardent zeal for some
   Resplendent form beguiles;
More lustrous than the dearest gem,
   True Beauty's essence smiles.

No image or analogy,
   By human mind or hand
Should lure our human fantasy
   To think we understand.

4. Too vast, too small, too far, too near:
   We cannot see God's face.
What thought can never comprehend,
   Yet love's strong arms embrace.

Eternity eludes our minds,
   And history shreds our hearts.
But he who called us to exist,

Sustaining grace imparts.

5. God loves us more than we, ourselves.
Divine compassion grieves
For all the pain and all the sin
That to our nature cleaves.

O, Holy, Holy, Holy Lord,
Omnipotent alone,
Who was and is and is to come,
We bow before your throne.

* * *

IN HOLY HEIGHTS

1. In holy heights still unrevealed,
Creator God enshrined,
You stood in majesty concealed
From human eyes or mind.

We could not look into the light
And dared not thus pretend
Your infinite, eternal might,
In creeds to comprehend;

2. Until, in Christ, omniscient plan
Was clothed in time and space,
And arms of flesh were drawn to span
The breadth of your embrace.

Now, closer than the breath of life
Or blood that in us flows,
You've entered into all our strife,
And hurt for all our woes.

3. Far more than we could love ourselves,
You've loved despite our sin,
And felt the fire of secret hells
We carried deep within.

You measured out upon a cross
Dimensions of your care,
On beams as broad as grief and loss,
And heavy as despair.

4. We love because you first loved us.
Your holy grace imparts
The substance and the impetus
Of love to human hearts.

Your everlasting arms embrace
Our guilt and self-disgust,
And dread of fate's gray, empty face
Is swept away in trust.